"This book has something for everyone—whether you're into conspiracy theories, religion, new earth, history, politics, or even you are looking for a safe place when 666 starts, this book will shock you"

Introduction:

I discovered a prophet from Argentina who, as early as the 1900s, made predictions that are now unfolding before us. The most startling aspect? Very few people know about him. Just five years ago, if anyone had suggested the possibility of UFOs, a Wall Street crash, a pandemic, or China preparing for something significant, it would likely have been dismissed as mere conspiracy theory. Concepts such as mixed genders and machines replacing human jobs seemed far-fetched. Yet, we

find ourselves witnessing these very scenarios come to life.

In this book, I will share with you the channeled drawings (psychographics) of Benjamin Solari Parravicini, each accompanied by the original Spanish explanation and an English translation. This will allow you to explore the material yourself, draw your own conclusions, and perhaps even notice details I might have overlooked. I feel a deep responsibility to bring this message to you.

My journey into Parravicini's world began when I noticed the buzz surrounding President Milei, who many were calling 'the Grey Man,' a figure supposedly fulfilling a nearly century-old prophecy. Intrigued, I delved deeper and discovered that this prophet had made *a thousand* predictions covering events across the globe. Remarkably, many of these prophecies appear to be unfolding before our very eyes.

Benjamin Solari Parravicini. He was a gentleman deeply rooted in the traditional society of Buenos Aires in the early 1900s. Initially, he channeled  voices to guide his work, but over time, his hand began to draw on its own, night after night, for years, often accompanied by writings next to his sketches. A fascinating twist in his story is the claim that he was abducted by aliens near the Buenos Aires Obelisk—an obelisk believed by some to be a conduit to other dimensions. While it may sound unbelievable, in esoteric circles, obelisks are considered significant.

Now, let's dive into the prophecies. I've added some illustrations to bring the predictions to life—some even coincidentally resemble Parravicini's own drawings. It seems as though there's a powerful force ensuring this message is communicated clearly. I believe it's our duty to share it with the world.

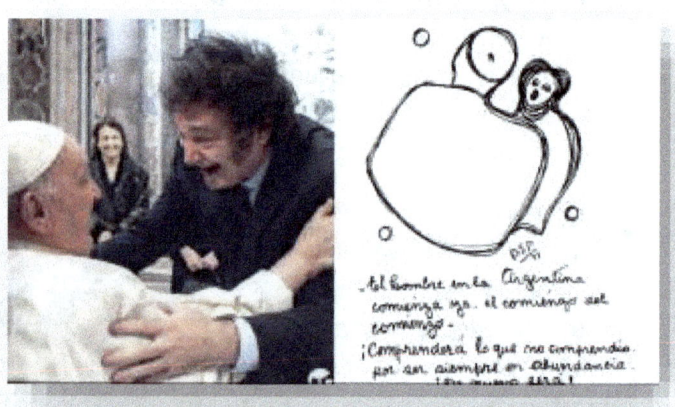

1) He predicted the TV **"Domestic Vision! You will see in your house through a small window external events"**

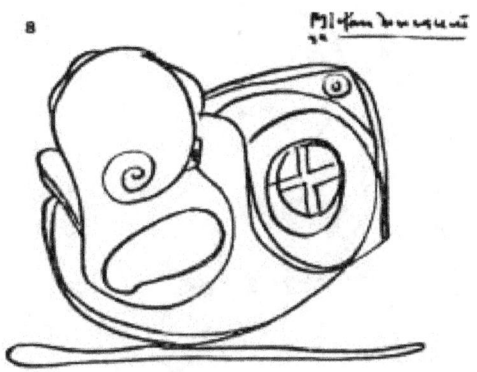

2) **"The dog will be the first to fly"**

El Can será el primer volador

3) **"Duel for power between Yankees (Americans) and Russians. Duel of space and land conquest. It doesn't look like it, but America will carry the Scepter".**

Russia seemed to be "winning" until Apollo 11 and the moon landing.

4) **"Arrives! A new telecommunications system through artificial planets".**
First satellites were round.

5) **"An American Golfer governs and he is killed young"**

6) **"The humble man in Argentina comes to govern. He will be of you unknown Caste and unknown in the environment, but he will be holy in manners, beliefs, and wisdom. He will arrive after the third day!"**. There are several prophecies regarding "The Grey Man", the man who takes Argentina out of socialism in its "French Revolution", which in victory, to the world. He talks about the "Caste" and won on the third round.

"The humble man in Argentina comes to govern. He will be of a young caste and unknown in the environment, but he will be holy in manners, beliefs and wisdom. He will arrive after the third day!"

PAX. El Hombre humilde en la Argentina se allega para gobernar. Él será de casta joven y desconocida en el ambiente, mas será santo de maneras, creencia y sabiduría. ¡Él llegará luego de la tercera jornada!

There are plenty of references to this Grey Man

7) **"Buenos Aires listens. Argentines, great! Military Will reign**". *This happened in the 70s.*

Buenos Aires you listen. Argentines, great! Military will govern

"Buenos Aires escuchas. Argentinos itino!. Gobernarán militares". (año 01/1939)

8) "The man of tomorrow will **emerge from laboratory sperm culture**. The woman will look to select the semen offered by the doctor. Preference will be given to athletic men (in reserve) and to intellectuals. The "vulgus" man wont be taken into consideration and they will let him disappear. The year 2000 after cataclysm will know this new norm and the animal man will have fallen asleep forever to give birth to men without original sin".

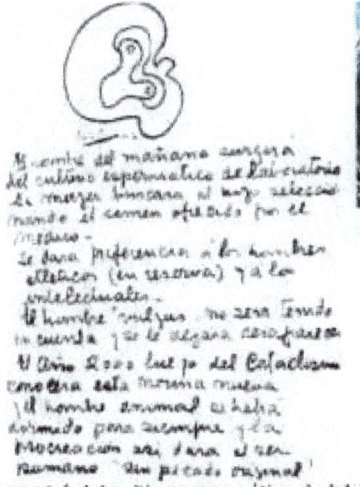

"El hombre del mañana surgirá del cultivo espermático de laboratorio. La mujer buscará al hijo seleccionando el semen ofrecido por el médico. Se dará preferencia a los hombres atléticos (en reserva) y a los intelectuales. El hombre "vulgus", no será tenido en cuenta y se le dejará desaparecer. El año 2000 luego del cataclismo conocerá esta norma nueva y el hombre animal se habrá dormido para siempre y la procreación así dará al ser humano "sin pecado original"". (año 1938)

9) **"The freedom of America will lose its light, its torch will not shine as it did yesterday, and the monument will be attacked twice"**. Twin towers and pentagon.

The freedom of America will lose its light, its torch will not shine as it did yesterday, and the monument will be attacked twice

10) Sexual contact will decrease.

- Men will reject **masculine women** and following absurd fashions.
- Women will be far from maternity die to the lack of desire of a **feminine man** and following absurd fashions.
- Laboratories will impose maternity artificially of the man of tomorrow and will be through sperm selection.

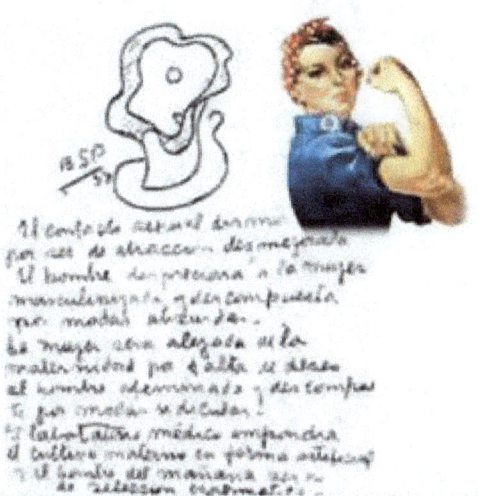

"El contacto sexual disminuirá por ser de atracción desmejorada. El hombre despreciará a la mujer masculinizada y descompuesta por modas absurdas. La mujer será alejada de la maternidad por falta de deseo al hombre afeminado y descompuesto por modas ridículas. El laboratorio médico impondrá el cultivo materno en forma artificial y el hombre del mañana será de selección espermática" (año 1937)

I'm sharing this next bit because, let's face it, our generation is becoming more uncommitted and individualistic by the day.

11) "Goodness will be pushed away by the constant laugh of the blind man that didn't listen. Darkness in him will be absolute and will roll towards infinite punishment. Pray!".

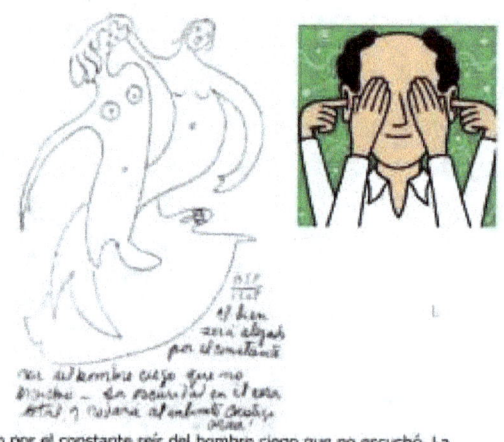

12) "Fire, hunger, pestilence, death, repeats the justice campaign that is approaching the world, even more so the world does not hear or see. Dragon comes the darkness that seemed asleep. The terror of bear who feigned love and brotherhood comes. **The humble democrat comes, who never was, and poverty comes with hum**. He, without shelter, and with them, all the explosions of disintegration".

Fire, hunger, pestilence, death, repeats the justice campaign that is approaching the world, even more so the world does not hear or see. Dragon comes the darkness seemed asleep. The terror of the bear who feigned love and brotherhood comes. The humble democrat comes, who never was, and poverty comes with him. He, without shelter, and with them, all the explosions of disintegration."

The "Yellows" and the "Reds" along the book it could also be read as libertarians and communism. After going through thousands of psychographics, I think he is talking about communism as "the storm". Read them by yourself and make your own conclusions!

13) **"The machine will displace/ replace humans. Humans will starve".**

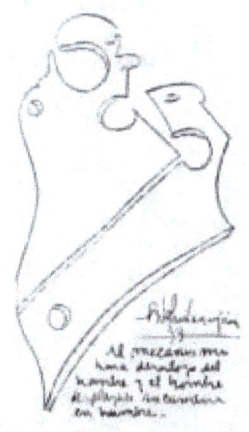

"El mecanismo hará desalojo del hombre y el hombre desplazado sucumbirá en hambre". Año 1939

14) **"Wars in the Middle East without rest will set the tone for the end of the finals in '66. In 66 man will want to die but no may, the explosion reaches the Antilles and..." (year 1938)**

"Wars in the Middle East without rest will set the tone for the end of the finals in '66. In 66 man will look in the mirror of the end and will want to die but not may, the explosion reaches the Antilles and..." (year 1938)

"El guerrear en Oriente sin descanso dará la pauta del final de finales el 66. El 66 el hombre se mirará en el espejo del final y querrá morir mas no podrá, la explosión llega a las Antillas y....". (año1938)

15) **"Death, pestilence and disasters will rise upon materialistic humanity, cataclysms. Year 65"**

"Death, pestilence and disasters will rise upon materialistic humanity. cataclysms. Year 65"

"Sobre la humanidad materialista se levantará la muerte, las pestes y los cataclismos. Año 65". (año1938)

16) **"No one in the world looks at the back of things, there <u>they prepare to invade</u>. Chaos arrives, it was said yesterday and today chaos arrives and Zion arrives"** *If I think of "the back of things" I think of "made in China". And "Zion" is also mentioned in the Matrix movies.*

"Nadie en el mundo observa el atrás de las cosas, allí se preparan para invadir. Llega el Caos se dijo ayer y hoy llega el Caos y llega Sion". (año1939)

17) **"The East will be fishing for opportunities, but they will become the fish".**

18) **"Extraterrestrial beings** will arrive on

earth again. They will arrive in different space ships... and they will inhabit the craters mountainous areas of the Andres and southern Patagonia. They will coexist human life, they will be seen and spoken to".

19) **"Beings will become visible to our retina** that travel on small fireballs, that penetrate houses and live in them... they are already on land".

20) The planets will tell man of non-humanoid races of **superior aesthetic beauty** and mental powers.

21) The ugly and deformed **monstrosity of "aliens" is human fantasy**, it does not exist!

22) CARIBBEAN. The yellows/ Russians / New Strength. Crown falls. Council.

New rules. "The alligator will bite and cause destruction. The yellow ones arrive. The reds will be scared because they will fear being victims of such a maneuver"

23) "The Satellites will fall per the yellow"
Relevant to cyber attacks? 1939?!

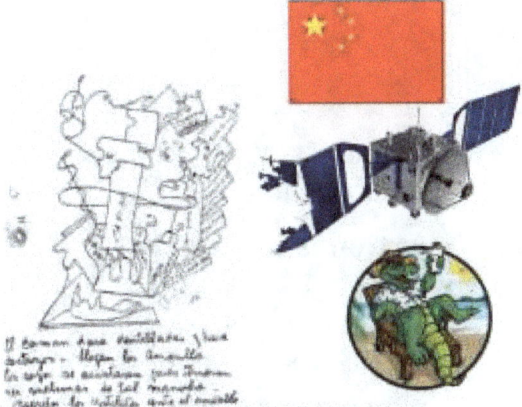

"CARIBBEAN. THE YELLOWS / RUSSIANS / NEW STRENGTH. CROWN FALLS. COUNCIL. NEW RULES. The **alligator** will bite and cause destruction. The yellow ones arrive. The reds They will be scared because they will fear being victims of such a maneuver. **·The satellites will fall per the yellow.**"
(year1939)

"CARIBE. LOS AMARILLOS / RUSOS / NUEVA FUERZA. CORONA CAE. CONCILIO. NUEVAS NORMAS. El caimán dará dentelladas y hará destrozos. Llegan los amarillos. Los rojos se asustarán pues temerán ser víctimas de tal maniobra. Caerán los satélites ante el amarillo".
(año1939)

24) **"New organisms will resurface as well as new insects, that been sleeping in the ice that is melting. 2000 will be again theirs. Humans by then will be brainless brains. Reality will be spooky because they will rule. Humans will escape".**

Didn't mention global warming, but talked about Poles Shifting, and this is the cause of natural disasters. Also man made weather modifications.

New organisms will resurface as well as new insects, that been sleeping in the **ice that is melting.** 2000 will be again theirs. Humans by then will be brainless brains. Reality will be spooky because they will rule. Humans will escape.

"Microbios nuevos saldrán a la palestra juntos con insectos nuevos. Surgirán animales antediluvianos, que durmieron en los hielos que cayeron en deshielo. El año 2000 será de nuevo de ellos. El hombre será para entonces de enormes cerebros más sin cerebros. La realidad asustará porque ellas serán las intelectas. El hombre escapará" (año 1934)

25) **"The Atom will take over the world. The world will be atomized and become blind. Storms will fall for messing with the atmosphere, new diseases, switched genders, social psychoses. The world will go dark"**.
Have you heard about HAARP? How they are modifying the weather?

"El átomo llegará a dominar el mundo, el mundo será atomizado y quedará ciego. Caerán tormentas ocasionadas por las incursiones del hombre en la atmósfera, nuevas enfermedades, trastoque de sexos, locura colectiva, dislate total. El mundo oscurecerá" Año 1934

26) "The lion of the seas falls. English power will be touched at its most intimate chord. <u>The communist comes to them</u>".

With the West increasingly leaning toward neo-

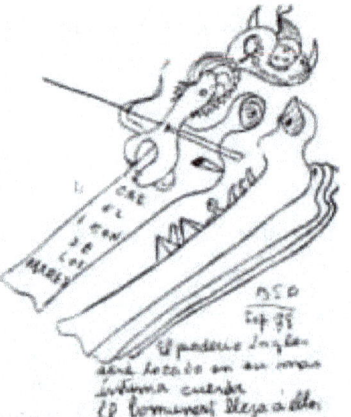

"THE LION OF THE SEAS **FALLS.** **English** power will be touched at its most intimate chord. The **communist comes** to them."

"CAE EL LEON DE LOS MARES. El poderío inglés será tocado en su más intima cuerda. El comunista llega a ellos". (año1938)

Marxism and leftist influences, it almost feels inevitable that they'll continue down this path. The push for globalism, a one-world government, Project Blue Beam, divisive rhetoric—all of these are steering these nations toward neo-Marxism. Meanwhile, South America and countries with past dictatorships seem to be waking up to what's happening.

There are plenty of references to the Fish-Man, to fish in general. We see this symbols all over the ancient world, in old maps, old books, and stories, and even today, the popes hat. For example, check this fish on the following sketch.

27) London!... the island is sinking"

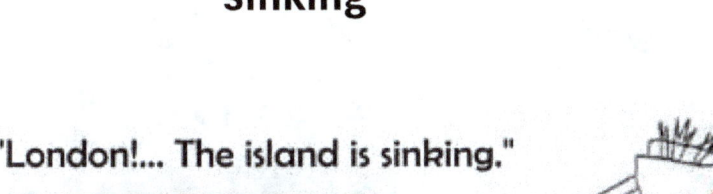

"Londres!... La isla se hunde". (año 1940)

28) **"It will fall with the cursed gold"**

"It will fall with the cursed gold."

"Caerá con el oro maldito". (año 01/1940)

29) South America. USA. North America. When the mountain range of South America trembles "Buenos Aires will tremble and then

COMMUNISM WILL BEGIN IN AMERICA".

SOUTH AMERICA. USA. NORTH AMERICA. When the mountain range of South America trembles "Buenos Aires will tremble and then **communism will begin in America**

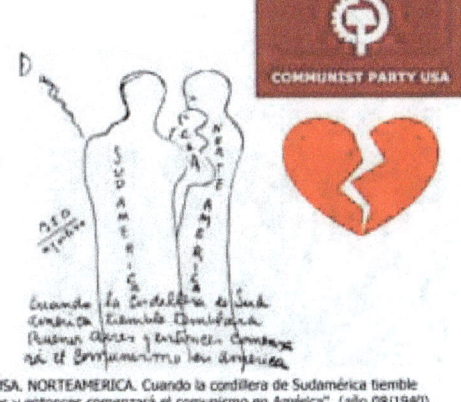

"SUDAMERICA. USA. NORTEAMERICA. Cuando la cordillera de Sudamérica tiemble temblará Buenos Aires y entonces comenzará el comunismo en América". (año 08/1940)

30) "Crisis – **stock market on the ground – the American millionaire will no longer be**".

Argentineans call Americans "Yankees".

Crisis - bolsa al suelo - el millonario yanqui dejará de ser (1940)

Norada RE

31) "Touristy **Cuba will be raised in disasters**, it will be the bear and the bear on its head it will remain for 5 times, then a host of surprises will surprise you. There will be blood, blood and fire, fire and death, and then nothing!" Hopefully the worst has happened.

32) "The center of the Americas will light up (fire?). The Yankees (Americans) will light up (burn?) and the Americans will turn against them. Caribbean will succumb".

"El centro de las Américas se encenderán. Los yanquis encenderán y las tomarán en su contra. Caribe sucumbirá". (año 1938)

33) "Argentina will have its "French revolution" in triumph, you can see blood on the streets if you don't see the moment of the "gray man".

Argentina was predicted to awaken from socialism, but it was warned that the country would descend into a bloodbath (like Venezuela) if we missed the moment of the "Grey Man" (which we didn't, thanks to Milei). Just look at the hole-hearted Cristina Fernandez de Kirchner, who Milei beat.

"Argentina will have its 'French revolution', in triumph, you can see blood on the streets if you don't see the moment of the 'gray man'".

"La Argentina tendrá su 'revolución francesa', en triunfo, puede ver sangre en las calles si no ve el instante del 'hombre gris'". (año 1941)

34) "Argentina will be in trouble without being there, because it heard what it was later understood and threw away. It will be in triumph!

I get that Argentina was in deep trouble, even without officially being communist—rising inflation, 60% poverty, and things just kept getting worse. But the people "heard" and later "understood," eventually rejecting the system, the socialist model. As we'll see, Argentina is poised to become the light of the world. Right now, we're just starting to listen because changing a culture takes time. But the impact of one man has been incredible. For the first time, Argentinians are being spoken to with numbers, statistics, and data. Until now, it's been all emotional speeches aimed at tugging at heartstrings, so we're in the midst of a cultural shift from word salads to cold, hard facts.

Argentina will be in trouble without being there, because it heard what it later understood and threw away. It will be in triumph!"

"Argentina estará en tropiezos sin estar, porque escuchó lo que luego comprendió y desechó. ¡Será en triunfo!". (año 1938)

35) The "middle class" saves Argentina. Its triumph will be in the world.

The drawings referencing George Orwell's Animal Farm, for illustration. The middle class has worked very hard for half of the population that didn't work, for many years.

"The 'middle class' saves Argentina. Its triumph will be in the world!"

"La 'clase media' salva a la Argentina. Su triunfo será en el imundo!". (año 1941)

36) Light, Russia, Japan, **"The Americas will bleed. Europe will bleed later. Every idea will see a sun of light, America will see the truth. Argentina will be light".**

Because Argentina has already been through the socialist system, they might be the ones to show the world how to break free from it. If Argentina starts improving, they could become a beacon of hope—a light for the world in this regard. Maybe Milei isn't the Grey Man, and maybe this isn't happening right now, but I can't shake the feeling that it is, and we're living through this prophecy. I began writing this book at the start of the year, and with each passing month, we seem to be inching closer and closer.

LUZ, RUSSIA, JAPAN, "The Americas will bleed. Europe will bleed later. Every idea will see a sun of light, America will see the truth. Argentina will be light."

LUZ, RUSIA, JAPON, "Las Américas sangrarán. Europa sangrará después. Cada idea verá un sol de luz, América verá la verdad. Argentina será luz". (año 07/1940)

37) "Argentina will suffer the storm in a small way, which will then hit the world. It will be example!".

"Argentina sufrirá la tormenta en pequeña, la que luego azotará al mundo. ¡Será ejemplo!". (año1938)

Again, I don't believe our storm was small—we faced significant poverty and struggle. But what he means is that we didn't descend into the depths of what happened in Cuba or Venezuela (though it was a very, very close call). The storm he refers to will then hit the rest of the world, with the U.S. and Europe turning communist. Argentina, in turn, will become the example of how to break free. We've already seen Milei leading by example at CPAC and other conferences.

Still, it's up to you to interpret the bio in your own way. When I first read it, I thought it was referring to an actual storm, but after reviewing all the slides, I realized what he was really talking about.

38) "Argentina, port of celestial door, of golden sands, of green pastures, of red flowers, will speak and say **'I have lands for those who suffer in burning, for those without a home for the orphaned child, for the hungry**, for the dispossessed, for the elderly, for the sick, for those who are born and for those who must be born in this place of promise. **Argentina Samaritan... of the world!"**

This wouldn't be the first time Argentina steps up in a big way. After the great wars, Argentina welcomed many Europeans with open arms and was even considered a world superpower back then. With its vast land, abundant resources, and a deeply charitable, community-focused population, Argentina has the potential to rise

"Argentina, port of celestial door, of golden sands, of green pastures, of flowers red, will speak and say: 'I have lands for those who suffer in burning, for those without home, for the orphaned child, for the hungry, for the dispossessed, for the elderly, for the sick, for those who are born and for those who must be born in this place of promise. Samaritan Argentina... of the world!

"Argentina, puerto de puerta celeste, de arenas de oro, de pastos verdes, de flores rojas, hablará y dirá: Tierras tengo para el que sufre en quemazón, para el sin hogar, para el huérfano niño, para el que hambre padece, para el desposeído, para el anciano, para el enfermo, para el que nace y para el que deba nacer en este lugar de promisión. ¡Argentina Samaritana... del mundo!". (año1942)

again. In this instance, he mentions the West being consumed by the "third smoke"—perhaps referring to a third world war?

39) "Absence or distraction will be a cause in Argentina in 66 grand shocks where the change of regimes will be known without bloody clashes but they will be a reality.

Argentina will suffer in small ways what the rest of the world will suffer later and the purified Argentina will be in the end of ends the SAMARITAN OF THE WORLD. Argentina understood. Argentina is rich because they have LOVE".

Argentina is community focused, very sociable and charitable society. If you would have told me four years ago, when most of the population was fanatized with the socialist movement, I would have laughed. I cannot believe this is a real possibility today.

Absence or distraction will be a cause in Argentina in 66 grand shocks where the change of regimes will be known without bloody clashes but they will be a reality. Argentina will suffer in small ways what the rest of the world will suffer later and the purified Argentina will be in the end of ends

the Samaritan of the world"

Argentina Understood.
Argentina is rich because they have LOVE.

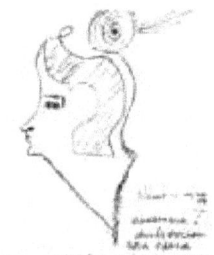

"Ausencia o distracción será causa en la Argentina en el 66 de grandes conmociones en donde se sabrá del cambio de regímenes sin choques cruentos pero serán una realidad. Argentina sufrirá en pequeño lo que el resto del mundo sufrirá luego y la Argentina purificada será en el final de finales la Samaritana de la Tierra". (año1938)

40) "Pilgrim, you who observe, go to the beach of the argentinean sands, there awaits peace. Lighthouse of lighthouses!".

This slide really struck a chord with me. It finally answered a question that's been lingering in my mind: **where will it be safe during the dark times ahead?** The answer seems to lie in Argentina—in places like the beaches, Neuquén, La Pampa, and other regions—where we're expected to receive new crops, supposedly thanks to help from astral beings, aliens, or some sort of interplanetary assistance.

But why will Argentina receive this help? Apparently, these beings will ask us to chant and comply, and Argentina will answer the call. There's a recurring emphasis on Argentina "being" or "becoming"—that verb appears over and over. And maybe, just maybe, it's because in Argentina, there's a unique sense of love.

"Pilgrim, you who observe, go to the beach of the silver sands, there awaits peace. Lighthouse of Lighthouses!"

"Caminante tu que observas, ve hacia a la playa de las arenas argentadas, allí aguarda la paz. ¡Faro de Faros!". (año1942)

41) "The song of the humble song will not be sung in the burning lands. It will be in Argentina, in union!".

42) "The Argentine is and will be in love because he knew how to sing and see God!"

"The song of the humble song will not be sung in the burning lands. It will be in Argentina In union!"

"The Argentine is and will be in love because he knew how to sing and see God!"

"El cantar del canto humilde no será cantado en las tierras en quemazón. Será en Argentina ¡En unión!". (año1942)

43) "Mexico, Cuba, Venezuela. Chaos".

Hopefully the worst passed.

"Méjico, Cuba, Venezuela. Caos". (año 1940)

44) "**Beings will be telepathic with the passage of time**. They already exist, and they already know their quality and hide it. The day will come when it will be in full light and will not surprise us".

For those that are into the New Earth, consciousness and energy, there is a lot of talk of developing psychic abilities once you increase your consciousness.

Beings will be **telepathic** with the passage of time. They already exist, and they already know their quality and hide it. "The day will come when it will be in full light and will not surprise us."

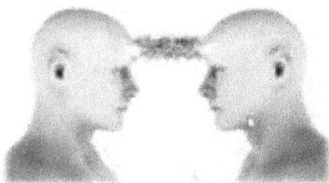

"Los seres serán telepáticos con el correr del tiempo. Existen ya, y ya saben de su cualidad y la esconden. Día llegará que será a toda luz y no asombrará". (año 1938)

45) "Darwyns theory will cease to exist because it will be known that the man came down from the planets".

We are the aliens! Following slides show how is it that humans are the aliens. And humans imagining aliens as beasts is human fantasy. They look just like us.

"Darwyn's Theory will cease to exist because it will be known that the man came down from the planets"

"La Teoría de Darwyn, dejará de ser porque se sabrá que el hombre planetas". (año sin fecha)

46) Epoque of **666**. The era of contradiction arrives".

I just googled a random symbol, but once you learn symbols you are going to see them everywhere. Nobody is going to be able to fool you. Even what happened in Haiti with Barbecue was a symbol.

"Época del 666. Llega la era de la contradicción". (año

47) "**666. BLIND WORLD.** The world in the seduction of evil will fall".

666. BLIND WORLD. The world in the seduction of evil will fall."

"666. MUNDO CIEGO. El mundo en seducción del mal caerá". (año sin fecha)

48) "Hunger will be with the yellow, brown and black finger that will destroy continents. The Angry exterminator will arise and the explosion of a thousand explosions will deafen and the sun will fall. Starts 66".

Would this be a weapon? A yellow, brown and black finger.

"**Hunger** will be with the yellow, brown and black finger that will destroy continents. The **Angry exterminator** will arise and the explosion of a **thousand explosions** will deafen and **the sun will fall**. Starts 66". (year without date)

"El Hambre será con el dedo amarillo, pardo y negro que destrozará continentes. El exterminador colérico surgirá y la explosión de mil explosiones ensordecerá y el sol caerá. Comienza el 66". (año sin fecha)

49) "Souls in the dark"

No Grid?

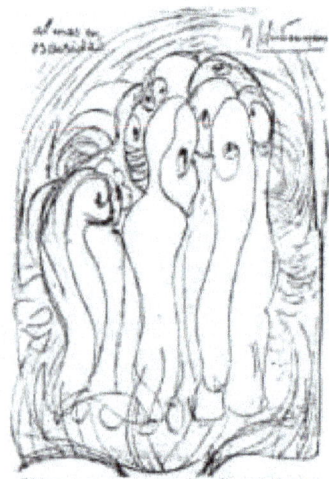

"Almas en oscuridad". Año sin fecha

50) "The world will go towards death since '66 for **not having known the vegetable brother, nor mineral, nor the thinker of the deep sea**, nor to the thinker of the **deep earth,** nor to the traveler interplanetary that will announce chaos and will not be understood".

We often focus on protein, but we overlook the importance of minerals.

The world will go towards death since '66 for not having known the **vegetable** brother, nor **mineral**, nor to the thinker of the **deep sea**, nor to the thinker of the **deep earth**, nor to the **traveler interplanetary** that will announce Chaos and **will not be understood"**

"¡Hacia a la muerte irá el mundo desde el 66 por no haber conocido al hermano vegetal, ni mineral, ni al pensante del profundo mar, ni al pensante de la profunda tierra, ni al viajero interplanetario que anunciará el Caos y no se le comprenderá". (año sin fecha)

51) "Two astronautical beings will arrive in Russia. Two astronautical beings will arrive to N.A. space beings will come to the world, they will be seen, and it will be said: they are the ones who "they came to populate the world yesterday".

To me, this represents the anti-Christ. It won't be an ordinary human. Consider the image of the Avatar above and the idea that Darwin's theory might fall when it's revealed that we originate from another planet. Just

Two astronautical beings will arrive in Russia. Two astronautical beings will arrive in NA Space beings will come to the world, they will be seen, and it will be said: they are the ones who "they came to populate the world yesterday."

"Dos seres astronáuticos llegarán a Rusia. Dos seres astronái N.A. Llegarán seres espaciales al mundo, se les verá, y se d llegaron a poblar en el ayer el mundo". (año193

piecing together the puzzle: interplanetary beings may have populated Earth long ago and resemble us. Additionally, the crystal ball with humans inside in the drawing hints at theories of Earth as a prison planet or a dome, concepts also found in the Bible.

52) "The world will crave the astronaut mask. They will bow and worship. The 'man-bird' will be admired, but this blindness will disappear when the thunder of war devastate the regions of the world"

When these beings arrive, the world will bow to them—or to him. However, this won't last, as the world will descend into devastation. We see similar motifs with the birdman in Egypt, the Mayans, and others. If you look closely at the

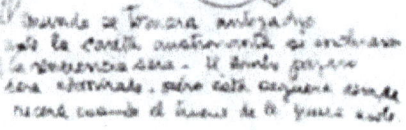

sketch, there's also another fish, which seems to be a recurring symbol.

Birdman- Sculpture by Sal...

53) "Astral thinking brains will come to earth and help you"

One good news! In another prophecy it says there are like stations next to the firmament of beings trying to help us too.

Astral thinking
brains will come
to earth and
help you

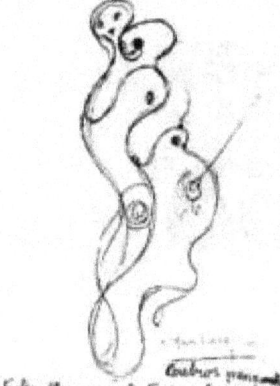

"Cerebros pensantes astrales llegarán a la tierra y le ayudarán". (año sin fecha)

54) "Noises! -**Noises in the bowels of the world.** -Noises in the coverage.- Noises in men. -Because it is the Hour 10 it is the punishment!"

Noises! -Noises in the bowels of the world.- Noises in the coverage. - Noises in men.-Because it is hour 10. In the punishment!

¡ Ruidos ! -Ruidos en la entraña del mundo.-Ruidos en la cobertura. -Ruidos en los hombres.-Porque es la hora 10. ¡ En el castigo ! (año 1972)

This is Julia Roberts in the Obama-produced movie "Leave the World Behind." You might be wondering what a politician is doing producing a Hollywood film. Anyway, this slide refers to when the world clock hits the 10th hour, marking our time for reckoning. Ancient clocks, like the astrolabe, reflect a flat earth model, with the moon indicating the month, stars the year, and the sun the time of day. It's quite a striking coincidence if you believe in the Big Bang theory. At the 10th hour of this world, it signifies Armageddon.

 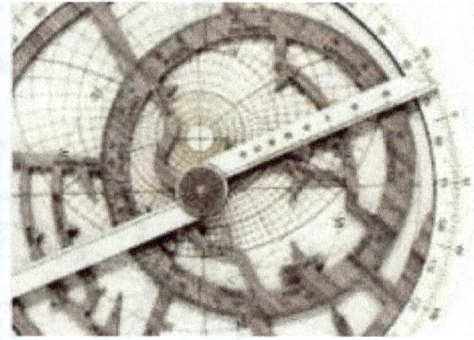

55) "Time marks the **end of the world**".

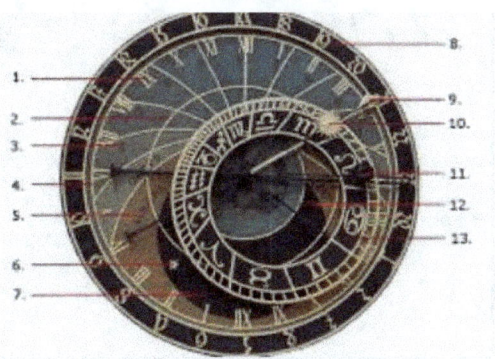

56) "Earthly civilization has a **lost civilization**. She was superior to those they followed. The man of that ime was "adaptable" to the great heat terrestrial, as to the great cold origin. He was electromagnetic, he knew how to govern himself by high mathematics and astral sciences. Still this magnetic man dwells frozen Pluto, because as said he is adaptable"

Conspiracy theories confirmation, Tartaria or Atlantis?

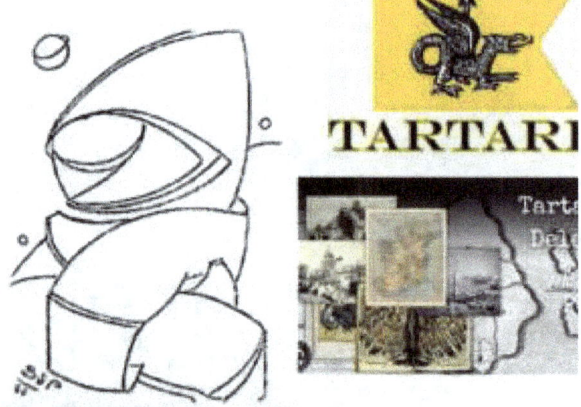

57) "The interplanetary amphibious ship will show **the existence of habitable channels in the interior of the Earth**. She will dive at the South Pole and appear at the North Pole, only in moments!"

The interplanetary amphibious ship will show the existence of habitable channels in the interior of the Earth. She will dive at the South Pole and appear at the North Pole, only in moments!

La nave anfibia interplanetaria enseñará la existencia de canales habitables en el interior de la Tierra. Ella se sumergirá en el Polo Sur y aparecerá en el Polo Norte, ¡solo en instantes! (año 1960)

Fish ship?

Which leads me to our next subject, **AI →**

58) "Brothers: cybernetics – **technological form of power will be murderous** of man upon arrival... of crying!"

I think what he's suggesting is that AI could develop the capability to harm humans once they start experiencing emotions. However, it doesn't necessarily mean it will lead to the end of humanity or that they will eliminate us all—just that they might have the potential for such actions. I imagine they could end up being like humans (or us, if we're not already like them), capable of love and connection, much like the characters in "Futurama" or the "Supersonics."

Brothers: Cybernetics - technological form of power will be murderous of man upon arrival.... Of crying!

Hermanos: La cibernetica - forma tecnológica de poder será asesina del hombre en el llegar.... ¡ Del llanto ! (año 1971)

59) **"Electronic music will be synced with dark music.** Both will desolate the world and the world will march towards the final chaos"

Electronic music will be synced with dark music. Both will desolate the world and the world will march toward the final chaos

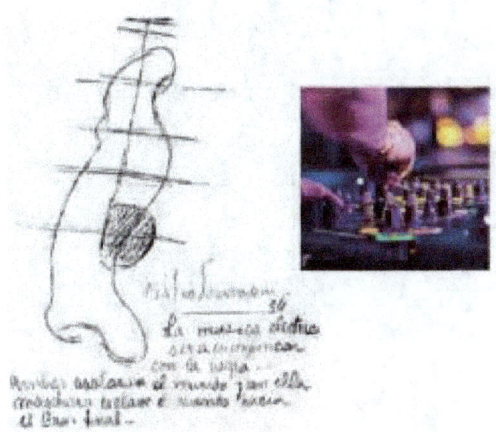

"La música eléctrica será en conjunción con la negra. Ambas asolarán el mundo y con ella marchará esclavo el mundo hacia el Caos final". Año 1936

I've heard they're changing the frequency from 432Hz to 440Hz, along with other theories about music. One thing I know for sure is that organic instruments are being replaced by computerized sounds.

60) "The uncapped desire for power will lose/ end the world"

The uncapped desire of power will lose/ end the world

"El deseo incontrolado del poder perderá al mundo". Año sin fecha

61) "Hunger. Charity needs to arrive in order to save the world".

62) **"Church will wrong their steps and their leaders will falsify it".**

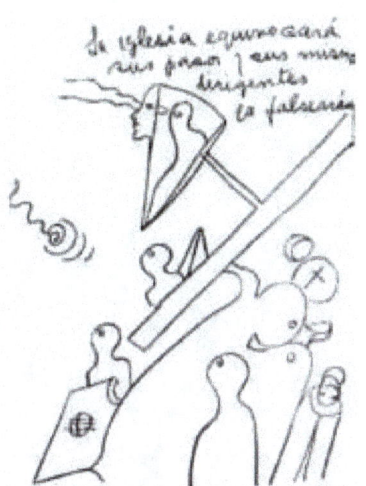

"La Iglesia equivocará sus pasos y sus mismos dirigentes la falsearán". (año sin fecha)

63) "The day will come the holy father will call priests to make them reason, but it will be in vain, the church will fall in blindness and despotism".

64) "Church, meditate! Your cloisters heal".

The day will come the holy father will call priests to make them reason, but it will be in vain, the church will fall in blindness and despotism.

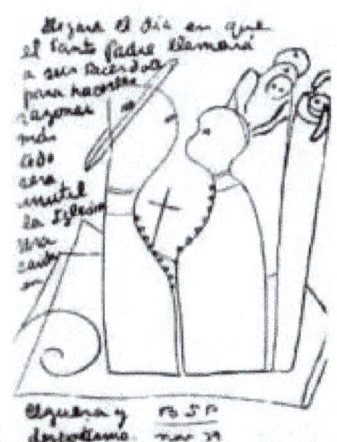

"Llegará el día en que el Santo Padre llamará a sus sacerdotes para hacerles razonar mas todo será inútil, la iglesia será caida en ceguera y despotismo". (año 11/1939)

That drawing 🙁

65) "With a dead pope the concordat of the Church will begin"

"Church Meditate! Your cloisters heal.

"With a dead pope the concordat of the Church will begin

"¡Iglesia Meditad! vuestros claustros curad". (año 1939)

I'm not diving into the church-related aspects here, but it's clear there's been a lot of hidden knowledge and guidance that the "above" has been trying to pass down. What really stands out to me is the **emphasis on meditation** and **chanting**. It's hard to believe this is just a coincidence—there's something significant about these practices that keeps coming up.

80) "Rome in misfortunes, the ducal city falls into disasters. Hermeticism in the neighborhood of Naples. Disorientation in the Vatican. The church is sinking, the pope will come out, it will become popular but late be. The reforms will scare catholics. Young priests will face currents power passers by in domain. New Church. New ways. Conciliations in fights. He tomorrow will be the return to the catacombs".

"Rome in misfortunes, the ducal city falls into disasters. Hermeticism in the Neighborhood of Naples. Disorientation in the Vatican. The church is sinking, the Pope will come out, it will become popular but late be. The reforms will scare Catholics. Young priests will face the currents power passers-by in domain. New Church. New ways. Conciliations in fights. He Tomorrow will be the return to the catacombs."

"Roma en desdichas, cae en desastres la ciudad ducal. Hermetismo en el Barrio de Nápoles. Desorientación en el Vaticano. La Iglesia se hunde, el Papa saldrá, se popularizará pero tarde será. Las reformas asustarán a los católicos. Los curas jóvenes enfrentarán a las corrientes pasatistas de poder en dominio. Nueva Iglesia. Nuevas formas. Conciliábulos en luchas. El mañana será el regreso a las catacumbas". (año 1938)

66) **"Kindness will disappear from the world. Robbery and crime will rule. Purity of beings will be corrupted by the bad example of unnatural homes. Marriages will decrease and lovers will be admired. Men will always allow seduction. They (Women) will be seducing them".**

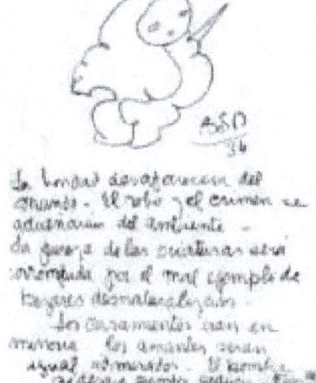

Kindness will disappear from the world. Robbery and crime will rule. Purity of beings will be corrupted by the bad example of unnatural homes. Marriages will decrease and lovers will be admired. Men will always allow seduction. They (women) will be seducing them.

"La bondad desaparecerá del mundo. El robo y el crimen se adueñarán del ambiente. La pureza de las criaturas será corrompida por el mal ejemplo de hogares desnaturalizados. Los casamientos irán en minoría, los amantes serán igual admirados. El hombre se dejará siempre seducir. Ellas serán las seductoras" (año 1934)

67) "Gonorrhea, diseases will return in '66 and wreak havoc"

"Gonorrhea, diseases will return in '66 and wreak havoc." (year without date)

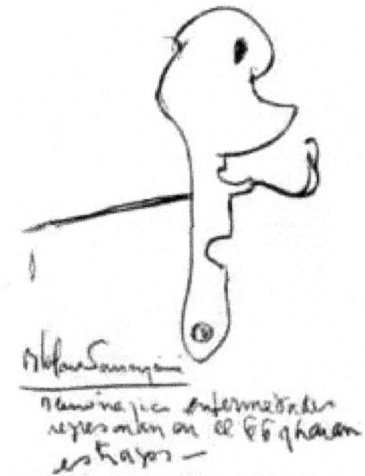

"Blenorragia, enfermedades regresarán en el 66 y harán estragos". (año sin fecha)

68) "Demonic era will be the era of the hour, hour of peace will come to this to demonstrate the existence of Christ. Egypt will say" The time, the clock, when we passed the 10 hour Christ will come back (could be a person, a consciousness, or something we don't really understand yet).

"Demonic era will be the era of the hour, hour of peace will come to this to demonstrate the existence of Christ. Egypt will say"

"Era demoniaca será la era de la hora, hora de paz llegará a esta para de la existencia de Cristo. Egipto dirá". (año1939)

69) Maranahata. Christ. Gospel. (Maranhata or maranhata means "Christ returns"). Means "Mary born" – "Mara Nahata".

Maranahata. Christ. Gospel."
(Maranhata or maranahata means
"Christ returns") (year without date)(means "Mary Born"-"Mara Nahata")

"Maranahata. Cristo. Evangelio". (Maranhata o maranahata si₍ vuelve") (año sin fecha)(significa "María Nacida"-"Mara N

Personal opinion alert. Take a moment to revisit Genesis, where God separates day from night, light from darkness—the son of God, the sun of God. Now, you might see Christ as a deity, a consciousness, or something we don't fully grasp. To me, the Bible is written in code (whether Jesus exists or not, it is a book of superior knowledge) that serves as the ultimate guidebook for this world. We're could be in a kind of prison, for the original sin, and the Bible could be the manual to help us transcend it.

We could think about the story of Jesus—not just as a tale of suffering, but as one of ascension. It's not about counting the blows or focusing on his torture. What truly matters is his resurrection—how he transcended the physical world. Jesus showed us the way, saying, "I am the way." It's not about who he was; it's about the message he delivered, the actions he demonstrated, and the path he paved. "I am who I am" means his name isn't what's important—it's the *how-to, the verb,* which matters. But that's just my interpretation. What's crucial is deeply understanding the message within these texts. That understanding, Parravicini channeled, is what will determine whether we truly get saved:

70) "Understanding of the holy scriptures will prevail at the end of time, it will be the north and salvation, it will be peace and unification of the Churches".

Understanding of the Holy Scriptures will prevail at the end of time It will be north and salvation, it will be Peace and unification of Churches".

"Comprensión de las Sagradas escrituras se impondrá al final de l será norte y salvación, será Paz y unificación de Iglesias". (año

Then we will see Mary, MA, as the moon. Electro-magnetic field, Sun = electric = male, moon = ma = magnetic= mother = female. Just theories! Theres also those who say its Virgo the sign.

71) "Faith. The world will go searching without seeing and seeing will not see, it will fall and rise with the Virgin".

Faith. The world will go searching without seeing and seeing will not see, it will fall and rise with the Virgin

"Fe. El mundo irá buscando sin ver y viendo no verá caerá y se la Virgen". (año 07/1939)

www.ingramcontent.com/pod-product-compliance
Lightning Source LLC
Chambersburg PA
CBHW071949210526
45479CB00003B/867